Understanding the Elements of the Periodic Table™

ARGON

Kristi Lew

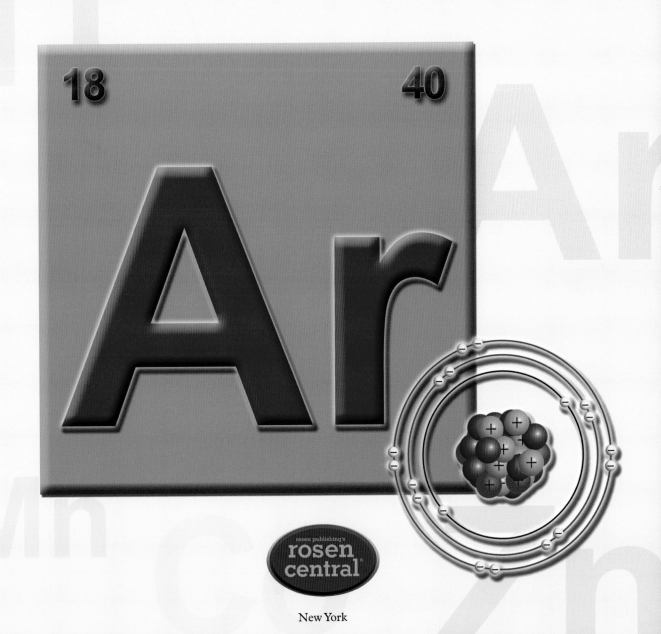

18   40

Ar

rosen publishing's
rosen
central®

New York

Published in 2008 by The Rosen Publishing Group, Inc.
29 East 21st Street, New York, NY 10010

Library of Congress Cataloging-in-Publication Data

Lew, Kristi.
Argon / Kristi Lew.—1st ed.
    p. cm.—(Understanding the elements of the periodic table)
Includes bibliographical references and index.
ISBN-13: 978-1-4042-1409-5 (hardcover)
1. Argon. 2. Chemical elements. 3. Periodic law—Tables. I. Title.
QD181.A6.L49 2008
546'.753—dc22

                                                    2007023213

Manufactured in Malaysia
On the cover: Argon's square on the periodic table of elements; *(inset)*
model of the subatomic structure of an argon atom.

Contents

Introduction

You cannot see, smell, or taste the element argon (chemical symbol: Ar). However, it surrounds you all the time. That is because argon is the third most abundant gas in the air. Seventy-eight percent of the air you breathe is made up of nitrogen (N_2, a molecule of two nitrogen atoms). At almost 21 percent, oxygen (O_2, a molecule of two oxygen atoms) makes up the next largest portion. That leaves about 1 percent not accounted for. Most of that 1 percent is argon.

Argon gas is constantly being made in tiny amounts in certain types of rocks. From those rocks, some of this argon gas slowly leaks into the atmosphere. Scientists separate the argon from other elements in the air so that it can be used. Electric lightbulbs and Geiger counters are usually filled with argon. (A Geiger counter is a device that is used to detect radioactive materials.) Argon is also used as a protective component for welding and growing silicon crystals. Argon has one important quality that makes it ideal for all of these applications: it does not react with most other chemicals, unlike the oxygen in the air that might otherwise be used.

Because argon is steadily being made by some types of rocks, scientists can use the relative amount of argon that is present to help them figure out how old those rocks are. This method of dating rocks works for volcanic rocks. It is called potassium-argon dating. Scientists date the rocks by comparing the amount of argon in the rock with the amount of a certain

These skeletal bones are from an ancient human, known as Omo I. Scientists used a form of potassium-argon dating to determine that the bones are approximately 195,000 years old.

type of potassium (K). Using this method, scientists can date rocks that were formed as long as four billion years ago!

In 1967, a scientist named Richard Leakey (1944–) led a team that discovered the fossilized bones of two ancient humans. The bones were buried under volcanic ash from an eruption that occurred a long time ago. Finding ancient bones is not that unusual for Leakey. He is a well-known paleoanthropologist. A paleoanthropologist is a scientist who studies protohuman (early human) and prehistoric human fossils.

Leakey called the fossils Omo I and Omo II because they were discovered near the Omo River in Kibish, Ethiopia. Shortly after they were discovered, these bones were dated to be about 130,000 years old. But, in 2005, almost forty years after Leakey found the bones, scientists determined that they are actually much older than that.

To figure out how old the bones were, scientists used a form of potassium-argon dating. They found that the Omo I and Omo II bones are actually 65,000 years older than Leakey believed. Today these bones have been dated at approximately 195,000 years old. They are the oldest bones of the modern human species, *Homo sapiens*, ever found.

Without methods such as potassium-argon dating, scientists would be limited in their ability to accurately date the rocks and fossils that they find. This method is a powerful tool for scientists who study rocks (geologists), dinosaurs (paleontologists), and ancient human life and culture (anthropologists).

Chapter One
Argon, It's in the Air

Before scientists could use argon to discover how old rocks were, though, they had to first discover argon. Because there are about 66 trillion tons (about 60 trillion metric tons, or 60×10^{12} metric tons) of argon in the air and it is the third most common gas in the atmosphere, it should have been easy to discover, right? Well, not exactly.

New Elements Are Hard to Find

In 1785, Henry Cavendish (1731–1810) was busy conducting experiments with air. He was trying to figure out what elements made up our atmosphere. He removed all of the nitrogen and oxygen from a sample of air by reacting them with other chemicals. Cavendish noted that every time he did this, about 1 percent of the atmospheric air he was studying did not react. But, that was as far as he got. Cavendish never did discover what made up the other 1 percent.

More than 100 years later, British physicist John William Strutt, commonly known as Lord Rayleigh (1842–1919), discovered something interesting while he was studying the density of nitrogen gas. The nitrogen seemed to have a different density depending on how it was obtained. Nitrogen isolated from the air was slightly denser than the nitrogen separated from the chemical compound ammonia (NH_3, a molecule containing one atom of nitrogen and three atoms of hydrogen). The difference was only about one-half of a percent, but instead of dismissing the conflicting data as experimental error, Rayleigh tried to figure out what was causing the densities to be different.

William Ramsay (1852–1916), a Scottish chemist, decided to help him investigate. Rayleigh had gotten the sample of nitrogen gas from the air by separating out all of the other gases that were known at that time. These gases were oxygen, carbon dioxide (CO_2, a molecule containing one atom of carbon and two atoms of oxygen), and water vapor (H_2O, a molecule containing two atoms of hydrogen and one atom of oxygen). When these three gases were extracted, Rayleigh believed the only leftover gas would

This photograph of Sir William Ramsay, seated on the left, and Lord Rayleigh was taken in 1894, the year they discovered the element argon.

be nitrogen. But, because of the discrepancy in the density data for this remaining portion, Rayleigh and Ramsay soon guessed that there might be some other, heavier gas still in the sample. This would account for the difference in density, they thought. They also hypothesized that this other gas must be nonreactive, or inert, because it had never been discovered before. Besides, it didn't react with the other known air components.

In 1894, Ramsay was finally able to separate the new, inert element. Instead of trying to isolate nitrogen from the air the same way Rayleigh had done, Ramsay changed the experiment a little bit. He used a different set of chemicals

Sir William Crookes used a portable spectroscope like this one to confirm that the gas Ramsay had isolated was, in fact, a new element.

to react and absorb not only all of the oxygen in the air, but all of the nitrogen, too. What was left over was the new, inert element.

Once Ramsay had the argon separated from the nitrogen, he asked another scientist, Sir William Crookes, to use a technique called emission spectroscopy to confirm that it was, indeed, a new element. Emission spectroscopy uses colored lines, called spectral lines, to distinguish one substance from another. Every element gives off a different combination of colored lines when it is heated, or, in the case of a gas, when electricity is passed through it. The light that is given off by the element is then separated into its component colors using a prism. When light from the sun is sent through a prism, the light is separated into a whole rainbow of colors.

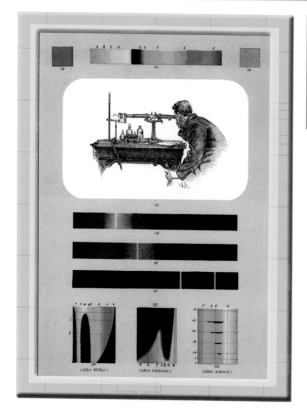

A spectroscopist looking through a spectroscope might see spectral lines like these. The number and colors of lines produced by a substance can help scientists identify that substance.

However, only part of the rainbow is visible when the light from a heated element is passed through a prism. The number and colors of spectral lines produced by an element are a type of signature that is distinctive of that element. When Ramsay and Rayleigh found argon for the first time, they knew they had found a new element because it emitted a set of spectral lines never seen before. Because the new element refused to combine with other elements, Ramsay and Rayleigh named it argon, after the Greek word *argos*, which means "lazy" or "inactive." Ramsay and Rayleigh announced their discovery in 1894.

The Periodic Table

The most common form of the periodic table used by chemists today arranges the chemical elements by increasing atomic number from left to right across the table. (See the periodic table on pages 38–39.) The columns are called groups or families and the rows are called periods.

Metals are generally found on the left-hand side and toward the bottom of the periodic table. Nonmetals are on the right side and toward the top. Metals in the middle, groups 3 through 12, are called the transition metals.

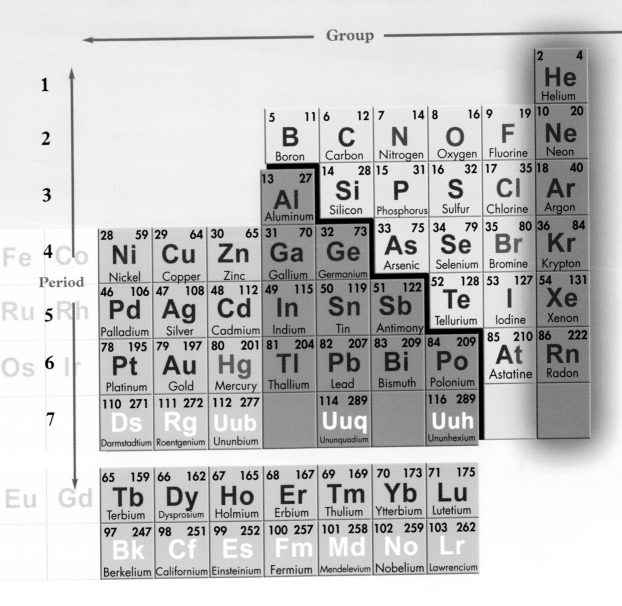

	VIIIB 10	IB 11	IIB 12	IIIA 13	IVA 14	VA 15	VIA 16	VIIA 17	O 18

Group

Period

1									2 4 He Helium
2				5 11 B Boron	6 12 C Carbon	7 14 N Nitrogen	8 16 O Oxygen	9 19 F Fluorine	10 20 Ne Neon
3				13 27 Al Aluminum	14 28 Si Silicon	15 31 P Phosphorus	16 32 S Sulfur	17 35 Cl Chlorine	18 40 Ar Argon
4	28 59 Ni Nickel	29 64 Cu Copper	30 65 Zn Zinc	31 70 Ga Gallium	32 73 Ge Germanium	33 75 As Arsenic	34 79 Se Selenium	35 80 Br Bromine	36 84 Kr Krypton
5	46 106 Pd Palladium	47 108 Ag Silver	48 112 Cd Cadmium	49 115 In Indium	50 119 Sn Tin	51 122 Sb Antimony	52 128 Te Tellurium	53 127 I Iodine	54 131 Xe Xenon
6	78 195 Pt Platinum	79 197 Au Gold	80 201 Hg Mercury	81 204 Tl Thallium	82 207 Pb Lead	83 209 Bi Bismuth	84 209 Po Polonium	85 210 At Astatine	86 222 Rn Radon
7	110 271 Ds Darmstadtium	111 272 Rg Roentgenium	112 277 Uub Ununbium		114 289 Uuq Ununquadium		116 289 Uuh Ununhexium		

n Fe Co
Ru Rh
Os Ir
m Eu Gd

65 159 Tb Terbium	66 162 Dy Dysprosium	67 165 Ho Holmium	68 167 Er Erbium	69 169 Tm Thulium	70 173 Yb Ytterbium	71 175 Lu Lutetium
97 247 Bk Berkelium	98 251 Cf Californium	99 252 Es Einsteinium	100 257 Fm Fermium	101 258 Md Mendelevium	102 259 No Nobelium	103 262 Lr Lawrencium

Part of the modern periodic table is shown here. Argon, one of the noble gases, is shown on the far right-hand side of the table in group 18 or O and in period 3.

The most commonly used modern periodic table has seven periods and eighteen groups. Each group is identified by a number, one through eighteen (an older system uses Roman numerals and capital letters IA–VIIA, IB–VIIB, and VIII that you still may find on a periodic table). Some groups also have common family names. Elements in group 1, for example, are called the alkali metals. The alkaline earth metals are in group 2 and the halogens are in group 17. Group 18 elements are known as the noble or inert gases. Elements found in the same group have similar chemical properties and sometimes similar physical properties.

A Whole New Group

When Ramsay and Rayleigh discovered argon, it was unclear how this element fit in with the other known elements. Because argon was the first inert gas to be isolated, and there were no other known elements with similar chemical properties, there was no obvious place in the periodic table for this new element.

Ramsay reasoned that this new element was probably not unique. Where there is one inert gas, there are likely to be others to make up a

What's in a Name?

Originally, argon's chemical symbol was "A." However, it was changed to "Ar" in 1957 by the International Union of Pure and Applied Chemistry (IUPAC), an international group dedicated to the advancement of chemistry. This change made argon's chemical symbol two letters like all of the other noble gas symbols. IUPAC has established a standard way of naming all the chemical elements and their compounds.

new group in the periodic table, he thought. He was right. Over the next several years, Ramsay would find several other inert gases.

Helium (He), the lightest of the noble gases, had been discovered in 1868 by Pierre Janssen (1824–1907), a French astronomer studying spectral lines emitted by the sun. But, Dmitry Mendeleyev, the Russian chemist who developed the first widely used periodic table, did not include helium in his 1869 version because not enough was known about the element. Helium was isolated on Earth for the first time by Ramsay and his assistant, Morris Travers (1872–1961), in 1895. They discovered that when they treated a particular type of rock, called cleveite (a uranium- and oxygen-containing ore), with an acid, the gas that was given off in the reaction produced the same pattern of spectral lines that were unique to helium.

After the discovery of argon and helium, Ramsay realized that there were probably other inert gases because families of elements usually contained more than two members. Three years later, in 1898, over a span of only forty-two days, Ramsay and Travers went on to find krypton (Kr), neon (Ne), and xenon (Xe). This extremely productive time followed two years of investigations that yielded few results toward this goal.

Ramsay suggested that maybe the periodic table needed to be extended to include a new group. Other chemists eventually agreed. Chemists decided to call the new group the noble or inert gases because they rarely bond with other elements. The last, and heaviest, noble gas, radon (Rn), was discovered by Friedrich Ernst Dorn (1848–1916), a German physicist, in 1900.

In 1904, Rayleigh won the Nobel Prize in physics for his part in the discovery of argon. Ramsay also won the Nobel Prize in chemistry that year for his discovery of most of the noble gases and his recognition of their placement in the periodic table.

Chapter Two
Atomic Argon

What makes the noble gases so inert? Argon and, indeed, all of the noble gases are so nonreactive because of their atomic structure. They all have what is known as a stable electron configuration.

Octet Rule

To understand why argon's electron configuration makes it stable, we should first look at how other elements combine with one another to form stable chemical compounds. The periodic table can, basically, be divided into two types of elements—metals and nonmetals. One of the typical chemical properties of nonmetals is that they gain electrons when they form compounds with metals. The number of electrons nonmetals will gain depends on how many electrons are in their highest energy level.

In a neutral atom, the number of protons, which have a positive charge, is equal to the number of electrons, which have a negative charge. This balance of charges is what makes the atom neutral. When a nonmetal atom gains an electron, it now has one more electron than the number of its protons. Consequently, it has a net negative charge.

If a metal atom gives up an electron, it has one less electron than the number of its protons. It now has a net positive charge. These charged

Protons, with positive charges, and neutrons, with no charge, are found in the center, or the nucleus, of an atom. Negatively charged electrons are in energy levels that surround the nucleus.

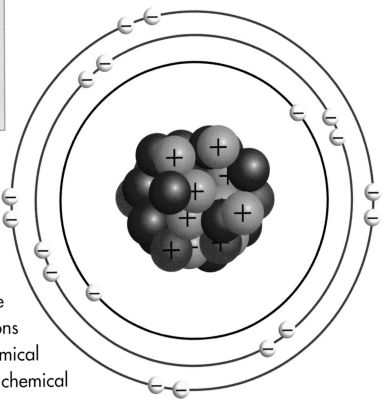

particles are called ions. When an atom loses or gains one or more electrons it becomes an ion. When a metal ion and a nonmetal ion come close together, the opposite charges of the ions attract one another and a chemical bond is formed. This kind of chemical bond is called an ionic bond.

The octet rule is a rule of thumb that says atoms of the s and p blocks (groups 1 and 2, plus 13 to 18) will tend to gain, lose, or share electrons to fill their outermost energy level with eight electrons (except for the lightest elements, such as hydrogen [H], helium, and lithium [Li], which are filled with two). (The s orbital is the lowest-energy orbital, and the p orbital is the next lowest-energy orbital, and each of these orbitals can hold a maximum of two electrons.) Electrons found in the outermost energy level (the highest-energy level) are called valence electrons. When a nonmetal forms an ionic bond with a metal, the metal gives up electrons and the nonmetal takes them so that both ions attain eight valence electrons in their outermost energy level that contains electrons.

For example, sodium (Na), an alkali metal, has an atomic number of eleven. This number means that a sodium atom has a total of eleven protons and eleven electrons. Up to two electrons fit in an atom's first energy level or shell. Eight total electrons can fit on the second energy

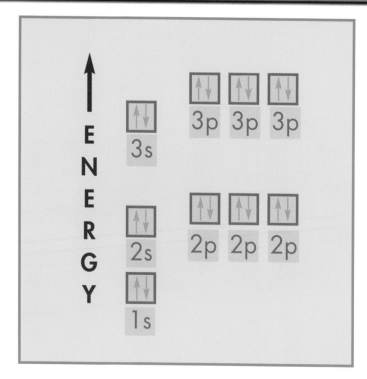

Argon's electron configuration shows its three energy levels and their sublevels. The first energy level has one sublevel, *s*. The second and third each contain two sublevels, *s* and *p*.

level. That leaves one electron in sodium that must go in the third energy level. According to the octet rule, and experimental evidence, sodium metal is reactive toward the loss of the single electron in the third shell. The octet rule predicts that an element will be most stable if it has eight electrons in its outer energy level that contains electrons. If the sodium atom loses the one electron in its third energy level, the second energy level now becomes the outermost energy level containing electrons. Because this level would now be full with its eight electrons, we would predict this to be a stable electron configuration by the octet rule. If the sodium atom lost this valence electron, it would become a sodium ion with a charge of positive one.

Chlorine (Cl), on the other hand, is a nonmetal. It has an atomic number of seventeen. A chlorine atom has seventeen electrons—two in the first energy level, eight in the second, and seven in the third. If chlorine could add one more electron to its third energy level, it would have a stable electron configuration because this shell would then contain eight electrons. This is exactly what happens when elemental sodium and chlorine react with each other to make sodium chloride. Sodium chloride (NaCl), better known as table salt, is formed when each atom of the metal sodium gives up one electron to an atom of the nonmetal chlorine. Because both

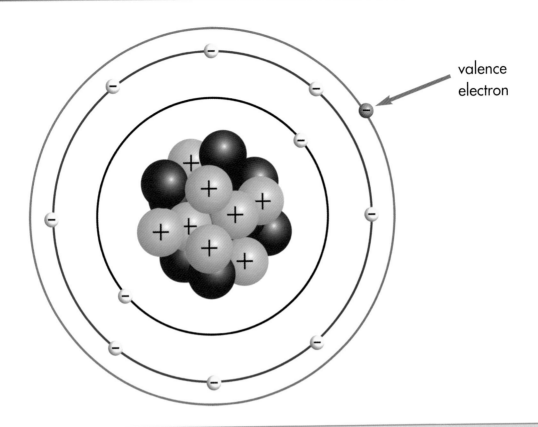

valence
electron

A sodium atom has one valence electron in its highest energy level. If this electron is lost, a sodium ion, with a charge of positive one, is formed.

elements end up with full outer energy levels, this forms a stable chemical compound.

Argon's atomic number is eighteen. So, an argon atom has eighteen protons and eighteen electrons. Two electrons will fit into the atom's first energy level, eight in the second, and eight in the third. Therefore, the third, and outermost energy level, is full, which is predicted to be stable according to the octet rule. This is why argon does not form compounds. Its electrons are already in a stable configuration. In other words, argon atoms do not need to react with other elements or compounds to pick up or give electrons away to attain eight electrons because they already have them. Full outer energy levels explain why the noble gases are stable elements that do not usually form compounds with other elements.

Argon Snapshot

18 40

Chemical Symbol:	Ar
Classification:	Noble or inert gas; group 18 of the periodic table
Properties:	Colorless, odorless gas; inert; not usually found in compounds
Discovered by:	Lord Rayleigh (John William Strutt) and Sir William Ramsay in 1894
Atomic Number:	18
Atomic Weight:	39.95 atomic mass units (amu)
Protons:	18
Electrons:	18
Neutrons:	22 (most common number is 22; however, argon has two other naturally occurring isotopes with 18 and 20 neutrons)
State of Matter at 68° Fahrenheit (20° Celsius):	Gas
Melting Point:	−308.83°F (−189.35°C)
Boiling Point:	−302.53°F (−185.85°C)
Commonly Found:	Third most common gas in the atmosphere

Isotopes of Argon

When Ramsay first suggested adding argon to the periodic table, there was some confusion because the element was simply a little too heavy. Because the elements at that time were arranged in the periodic table by increasing atomic weight (today they are arranged by atomic number), argon's weight should have been between chlorine and potassium. However, argon's average atomic weight is 39.95 atomic mass units (amu), while potassium's is only 39.10 amu.

Later, other scientists discovered that the reason argon is heavier than potassium has to do with the relative amounts of the two element's isotopes. Isotopes are forms of the same element that have different mass numbers. They have the same number of protons but different masses because they have different numbers of neutrons. Argon's major isotope is argon-40. The number behind the element's name (which is also sometimes shown to the right as a superscripted number) shows the mass of the isotope. An argon-40 atom has eighteen protons, as all argon atoms have, and twenty-two neutrons. The masses of protons and neutrons are nearly equal to each other and to 1.0 amu, so the expected mass for an atom is

Not So Noble After All

A compound of argon was actually made in the year 2000. Along with argon, the compound contains the elements fluorine (F) and hydrogen (H). It is called argon fluorohydride, or HArF. It is not a very stable compound, however. In fact, it falls apart into argon and hydrogen fluoride (HF) unless it is kept extremely cold—below −411° Fahrenheit (−246° Celsius).

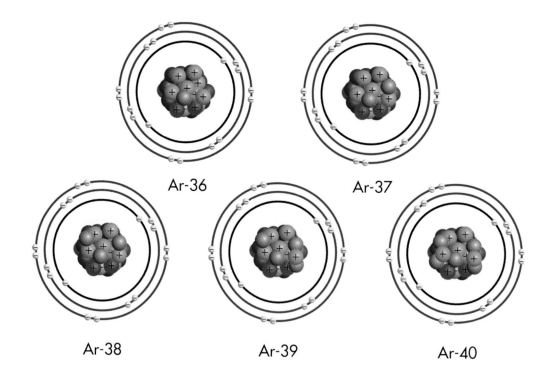

Ar-36   Ar-37

Ar-38   Ar-39   Ar-40

The isotopes of argon all have eighteen protons and eighteen electrons. Each isotope has a different mass number because each contains a different number of neutrons.

simply determined by adding the number of protons and the number of neutrons together. The mass of an electron is much, much smaller than (less than one-thousandth) the mass of a proton or neutron and is ignored when estimating the atomic masses this way. There are also an argon-38 isotope and an argon-36 isotope containing twenty and eighteen neutrons, respectively. But about 99.6 percent of the argon found in the atmosphere is argon-40.

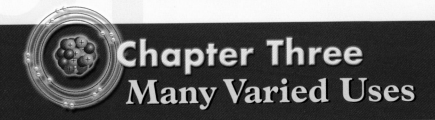

Chapter Three
Many Varied Uses

Particular isotopes of potassium and argon are what allow scientists to date artifacts such as the Omo I and Omo II bones. Potassium-argon dating relies on the fact that certain isotopes of potassium break down into argon atoms.

Potassium-Argon Dating

Potassium-39 is one of the most abundant isotopes in the earth's crust. This isotope of potassium is stable. But, about one in every 10,000 potassium atoms is a different isotope that is not stable. It is radioactive.

A radioactive element has an unstable nucleus that eventually reacts by a process called radioactive decay. During radioactive decay, the nucleus of the unstable element undergoes a change. This transformation in the nucleus often changes the number of protons in the unstable element to make a different element out of the atom. Because the number of protons in the nucleus of an atom defines the identity of each element, every atom of a particular element has the same number of protons. That means that all isotopes of an element have the same number of protons in their nuclei. Otherwise, it would be a different element entirely.

Scientists can use the potassium-argon dating process to accurately date volcanic rock that formed as long as four billion years ago.

All argon atoms have eighteen protons. Eighteen protons in the nucleus of an atom are what make that atom argon and not some other element. When a radioactive atom decays, though, the number of protons in its nucleus can change, transforming the element into a different element (which may or may not be more stable). The radioactive isotope of potassium is potassium-40. Radioactive decay causes potassium-40 to change into argon-40. This is a natural process where an atom of potassium-40, which is called the parent, undergoes a nuclear reaction. This particular nuclear reaction forms an atom of argon-40, which is called the daughter.

There are several different types of radioactive decay reactions. The type that potassium-40 undergoes to change into argon-40 is called electron capture. In electron capture, one of the electrons closest to the nucleus "falls" into the nucleus of the atom. When an electron is captured by a potassium-40 nucleus, the electron combines with a proton to form a neutron and another subatomic particle called a neutrino. The neutrino, which is a neutral particle of a very small mass, is ejected from the nucleus. Because the atom loses a proton during this process, it is changed into a different element. When a potassium-40 atom loses a proton, it is changed into argon-40.

$$^{40}_{19}\text{K} + {}^{0}_{-1}\text{e} \xrightarrow{\text{electron capture}} {}^{40}_{18}\text{Ar} + {}^{0}_{0}\nu$$

Through electron capture, a potassium atom, atomic number 19, becomes an argon atom, atomic number 18. The numbers in front of the element's symbol represent that element's atomic and mass numbers. The numbers on the bottom are atomic numbers. The numbers on top are mass numbers. The symbol "e" stands for an electron. The number on the bottom is the charge of the electron. Electrons have no mass, so the number on top is zero. The neutrino is represented by the Greek letter nu (ν).

Using a radioactive isotope like potassium-40 to date artifacts is called radiometric, or radioactive, dating. The method relies on the fact that potassium-40 is radioactive. If potassium-40 atoms were stable instead of radioactive, the process would not work. That is because as the potassium-40 isotopes decay, they change into argon-40 in a specific way. This particular change always happens the same way (through electron capture), and the time that this process takes stays constant and is known. Consequently, scientists can use the amount of time necessary to turn potassium-40 atoms into argon-40 as a "clock" to determine how old something is.

The rate at which a radioactive isotope decays is called its half-life. In potassium-argon dating, scientists use the half-life of potassium-40 to help them figure out the age of the rocks. The half-life of a radioactive element is the amount of time it takes for half of the radioactive parent atoms to decay into the daughter isotope(s). Potassium-40 has a half-life of 1.25

billion years. That means that it would take 1.25 billion years for half of a sample of radioactive potassium-40 atoms to undergo nuclear reactions. Potassium-40's extremely long half-life is what makes this radioactive isotope a good choice for dating very old rocks.

Radioactive potassium-40 isotopes are found mostly in igneous, or volcanic, rocks. When molten lava from a volcano solidifies, the radioactive countdown begins.

When scientists use potassium-argon dating, they assume that when the rock is formed by the cooling of magma that no argon-40 is present. The naturally occurring potassium-40 is present, however, and over time slowly decays to form some argon-40, which is trapped inside the solid rock. To get an accurate date, scientists must heat the rock to release the argon-40 trapped inside. Then they measure how much argon-40 and potassium-40 are in the rock sample. This measurement tells them what percentage of potassium-40 atoms have decayed into argon-40, and then they can determine the amount of time that was required to do so.

There are a few problems with potassium-argon dating, however. One problem occurs if the volcanic rock is reheated after being formed. If this happens, some argon may leak out of the rock. Because some of the argon has escaped, it would appear as if the rock was younger than it actually is. Reheating and cooling "resets" the potassium-argon clock.

Sometimes excess argon-40 can contaminate a rock and make it appear older than it actually is. This happens if a rock beneath the one being dated has melted and released some argon-40 that has seeped into the rocks above. This excess argon-40 is called motherless argon-40 because the "mother" potassium was never in the sample being dated. That would mean that the argon-40 that is produced does not actually belong to the sample being tested. When a sample contains more argon than it should, it makes scientists think that the rock is older than it really is. Experts using this method select their samples carefully to minimize these errors and run the dating on multiple samples to increase the reliability of their results.

Potassium-argon dating is used to date ancient volcanic rocks. Here, Kilauea, a volcano on the island of Hawaii, produces lava that will eventually solidify into new volcanic rock.

Argon-Argon Dating

Argon-argon dating is a similar method for dating rocks. For this technique, different isotopes are used—argon-39 and argon-40. Argon-39 is not a naturally occurring isotope. It is a radioisotope with a relatively short half-life of 256 years. Scientists can make argon-39 in the laboratory by exposing a sample of rock to a nuclear reactor.

When the rock sample is placed in the nuclear reactor, it causes the potassium-40 in the rock to form argon-39. This is not a natural process. Remember that in nature potassium-40 breaks down into argon-40, not argon-39. So, now the rock sample to be dated contains not only the argon-40 that potassium-40 naturally decayed into, but also argon-39 that the scientists have produced using the nuclear reactor. The rock is next heated by a laser to release both the argon-40 and the argon-39 gases. Based on the amount of argon-39 released, the scientists can calculate the amount of potassium-40 that was originally in the rock. This calculation helps the scientists determine the rock's age. The advantage of this technique over potassium-argon dating is that only one measurement is made, the ratio of argon-40 to argon-39. This avoids having to measure the potassium-40 amount in addition to the argon-40 amount. Besides its use in dating rocks, argon has applications in many industrial processes.

Welding

Argon's main purpose in industry is in the welding process. When two metals are welded together, they are heated to extremely high temperatures. Eventually, the metals get hot enough to melt together. This procedure forms a very strong joint. However, when metals get hot, they react with oxygen very easily and start to form compounds with the oxygen. This

Argon is used to shield metals from the oxygen in the air during the welding process, allowing them to weld correctly.

type of reaction is called oxidation. If either of the metals being welded already has been oxidized, it is hard for them to join together. And, any welds that are made will be weak and more likely to fail. Think of how rust, which is oxidized iron, is not strong and how easily it can flake away. So, welders use inert gases, such as argon, to serve as a "blanket," or shield, for the welding process. One of the properties of argon that makes it especially useful for this purpose is that it is more dense than air. When a welder pulls the trigger on a gun attached to the welding machine, argon flows from a tank, through a tube, and out of the end of the tube to cover the metals being welded. The flow of argon pushes the air, containing oxygen, out of the way so that the metals can weld correctly. Argon is often used in arc welding metals such as aluminum (Al), stainless steel (which is iron [Fe], carbon [C], and chromium [Cr]), magnesium (Mg), and titanium (Ti).

Steel Manufacturing

Steel is an alloy. An alloy is a mixture of two or more elements. At least one of the elements in an alloy is a metal, and alloys are usually made to improve the properties of the metal. Steel can be made by mixing iron with carbon. Other elements, such as chromium (Cr), manganese (Mn), or tungsten (W), also may be added. Steel is much stronger than iron is without the carbon and other elements mixed in.

When stainless steel is made, argon and small, controlled amounts of oxygen are bubbled through a mixture of melted regular steel and chromium. The oxygen reacts with carbon to make carbon dioxide. This decreases the amount of carbon in the steel, as stainless steel contains a lower carbon percentage than regular steel. Because the argon will not react with any of the ingredients used to make the stainless steel, it is employed to keep the mixture bubbling and mixed. The resulting stainless steel is strong and corrosion resistant, which means that the iron in it does not usually react with oxygen or other chemicals in the environment that would cause tarnishing or rust formation. It is the chromium in the stainless steel that protects the iron. The way that the chromium does this is to react with oxygen at the surface to make a protective, thin, chromium oxide layer that is invisible, strong, and adherent. It is this layer that keeps the oxygen and water away from the iron underneath it to prevent the iron from oxidizing or rusting. This makes stainless steel a very useful alloy for making all sorts of metal objects, including pots and pans, knives, garden equipment, outdoor furniture, barbeques, and even the kitchen sink!

Electronics Industry

The semiconductor industry uses argon for growing silicon (Si) and germanium (Ge) crystals that are very important in the computer, cell phone, and digital audio industries. The argon surrounding the crystals is nonreactive and keeps other potential contaminants away from the surface of the crystal while it is forming. This leads to purer crystals that make better electronic equipment.

Geiger Counters

A Geiger counter is a device used to detect radioactive materials. Radioactive substances give off radiation in the form of energy or energetic particles, which are produced by reactions of the nucleus. There are two classifications for radiation—ionizing and non-ionizing. Non-ionizing radiation can have enough energy to move atoms around, but it is not powerful enough to change them chemically. Visible light, sound waves, and microwaves are all examples of non-ionizing radiation.

Ionizing radiation, on the other hand, is very powerful. This type of radiation has enough energy to remove electrons from atoms to form ions. Types of ionizing radiation produced in nuclear reactions include particles such as alpha particles (which are two protons and two neutrons bound together) and electrons (also called beta particles), or gamma rays (these are packets of energetic electromagnetic radiation or light, sometimes called photons). Ultraviolet light from the sun is a natural source of radiation that can cause sunburns and break chemical bonds. X-ray tubes and nuclear power plants are artificial sources of ionizing radiation.

Geiger counters are very sensitive instruments that can measure even low levels of ionizing radiation. Argon is often used to fill Geiger counters.

Argon is sometimes used to fill a Geiger counter, a device that can detect dangerous ionizing radiation.

When an energetic particle or a form of light such as an alpha particle, a beta particle, or a gamma ray enters the Geiger counter and encounters an argon atom, the argon atom loses an electron. This creates an argon ion. The ionized gas is now able to conduct electricity. Before the electrons were knocked off, it could not. The Geiger counter indicates a pulse of electricity that is formed by this process with a sound such as a click. The more clicks there are, the more argon is being ionized, and the larger the quantity of ionizing radiation there is entering the window of the Geiger counter.

Argon is not solely used in industry. It has many varied applications—from lighting to health care to preservation.

Lights, Lights, Lights!

The main use for argon around the house is to fill incandescent lightbulbs. An incandescent material is one that gives off light when it gets hot. In the common lightbulb, that material is a filament made of tungsten metal. The filament is the thin wire inside the bulb that stretches across the middle of the lightbulb. When a light switch is turned on, electricity flows through the filament, heating it up to around 4,000°F (2,200°C). Tungsten is used for the filament because, unlike most other metals, tungsten does not melt at these extremely high temperatures. But, even though the tungsten does not melt, it can combust (catch on fire) or evaporate.

Combustion, which is another word for oxidation, requires oxygen. If lightbulbs were filled with air, the oxygen in the air could cause the tungsten to catch on fire. To prevent this from happening, early lightbulb manufacturers sucked all the air out of the lightbulbs, creating a vacuum—an area with no matter, not even a gas. The problem with this approach is that at these high temperatures, atoms of tungsten sometimes evaporate

Incandescent lightbulbs are filled with argon gas because it does not react with the tungsten metal filament (glowing wire). Argon also slows the evaporation of tungsten atoms from the filament.

from the filament. This causes the filament to get thinner and thinner and, eventually, the filament would break and the lightbulb would no longer work.

Another problem with atoms of tungsten evaporating from the filament was that the tungsten atoms were being deposited on the inside of the lightbulb, making it black. The evaporation of tungsten from the filament shortens the life of a lightbulb considerably. Evaporation occurs more readily in a bulb under a vacuum compared to one filled with a gas.

In the early 1900s, scientist Irving Langmuir (1881–1957), while working for the General Electric (GE) Research Laboratory, suggested that GE fill lightbulbs with an inert gas. Today the inert gas that fills most light-bulbs is argon. Argon will not react with the tungsten like oxygen does. In addition, filling the bulb with a gas substantially decreases the evaporation rate of tungsten atoms from the filament. As a result, filling the bulbs with argon solved many of the problems early lightbulb manufacturers faced.

Windows

Lightbulbs are not the only place argon is used in the home. Double-paned thermal windows, which have two panes of glass separated by a space,

Irving Langmuir, shown here in 1940 at his microscope in the General Electric Research Laboratory, suggested that GE fill its lightbulbs with an inert gas, such as argon.

are also filled with argon. Argon is used instead of air to fill the space between the panes of glass because argon does not conduct heat as well as air does. As a result, argon-filled windows lose less heat or are more insulating than double-paned windows filled with air.

Argon and Health Care

The blue-green light that argon lasers emit is especially suited to treating eye diseases. So far, scientists have discovered that argon lasers are useful in the treatment of several eye disorders such as some forms of glaucoma and age-related macular degeneration (AMD). The blue-green light of the argon laser has wavelengths between 455 and 529 nanometers (nm; a

"Neon" Lights

Argon is also used in "neon" signs. Neon produces a reddish orange color when electricity is discharged through it. Consequently, if only the element neon is used, all neon lights would be red! Instead, several elements besides neon are used in these lights to produce different colors. When the light tubes are filled with argon, the light glows light bluish purple. If a little mercury (Hg) is mixed with the argon, a brighter, deeper blue is given off. For yellow lights, helium is used to fill the tubes. Tubes filled with krypton are a greenish gray color, and xenon glows blue-gray.

This neon sign is reddish orange because the tube is filled with neon gas. If argon is used instead, the letters would be a bluish purple color.

nanometer is one billionth of a meter and is used in the measurement of light wavelength), with dominate emissions at 488 nm (blue) and 514 nm (green). The wavelength of visible light determines its color and the amount of energy that it contains. The energy from the light rays at these wavelengths destroys abnormal blood vessels in the eye with heat. Before using argon lasers, doctors tried lasers that emit red light. At around 650 nm, red light has the longest wavelength of visible light and the least amount of energy. Because of this difference in wavelength and the lesser amount of energy contained by red light, the red lasers did not correct the eye disorders as well as the argon lasers.

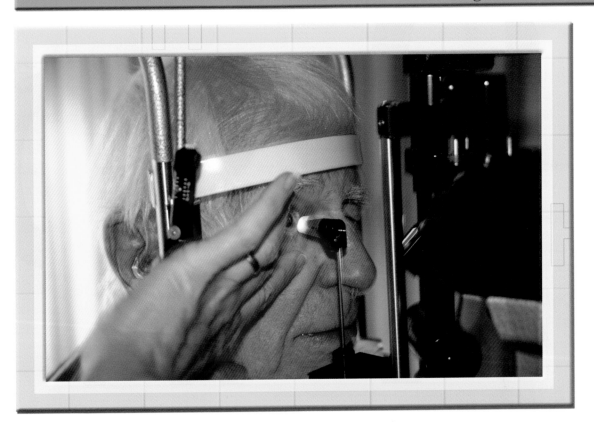

Argon lasers can be used to treat a variety of eye disorders, including some types of glaucoma and age-related macular degeneration. They can also be used in PRK, a corrective eye surgery.

Argon lasers are also used in corrective eye surgery. Fluorine gas is mixed with the argon gas in a laser system that produces a beam of ultraviolet light. This type of laser is used in photorefractive keratectomy (PRK), a procedure similar to LASIK. (In LASIK surgery, a laser is used to create a flap in the cornea of the eye, to remove tissue and to reshape the cornea so that focus power is changed.) The fluorine/argon laser beam is special because it can remove one microscopic layer of tissue at a time from the cornea without harming the healthy tissue around it. When the energy from the laser beam hits the cornea, it vaporizes the tissue by breaking the bonds between the molecules in the eye. Every pulse of the laser takes off about 0.4 microns of tissue. For comparison, one human hair is between 50 and 100 microns thick. These lasers are very accurate.

Argon is also used in cryosurgery on the body. During cryosurgery, diseased tissue or abnormal cells, such as cancer cells, are destroyed using extremely cold temperatures. Liquid nitrogen is usually used to achieve such cold temperatures if the tissue is on the outside of the body. When the tissue is internal, either liquid nitrogen or argon is used. A long, thin wandlike device, called a cryoprobe, is used to get to the diseased tissue. Liquid argon (−302°F, or −186°C) or liquid nitrogen (−320°F, or −196°C) is pumped through the probe to super-cool the probe tip. Temperatures at or below about −40°F (−40°C) kill cells.

The Declaration of Independence is housed in a gold-plated titanium case filled with argon, which will not react with the paper's fibers and, therefore, will not destroy it like oxygen would.

Charters of Freedom

Because of argon's inert nature, the gas is good for preserving things that would normally disintegrate if they were exposed to the oxygen and/or water from air. The United States' Charters of Freedom, which include precious national documents such as the Declaration of Independence, the U.S. Constitution, and the Bill of Rights, are being protected by argon. All of these documents are enclosed in cases filled with argon that will not react with the papers' fibers like the oxygen in air would. If you have seen old paper that has turned brown, you have seen what oxygen does to paper; basically, it is just a very slow oxidation or combustion process.

From welding, to health care, to filling lightbulbs, to preserving precious national documents, argon has one special chemical property that makes all of these uses possible—it is inert. In this case, being "lazy" is a good thing!

The Periodic Table of Elements

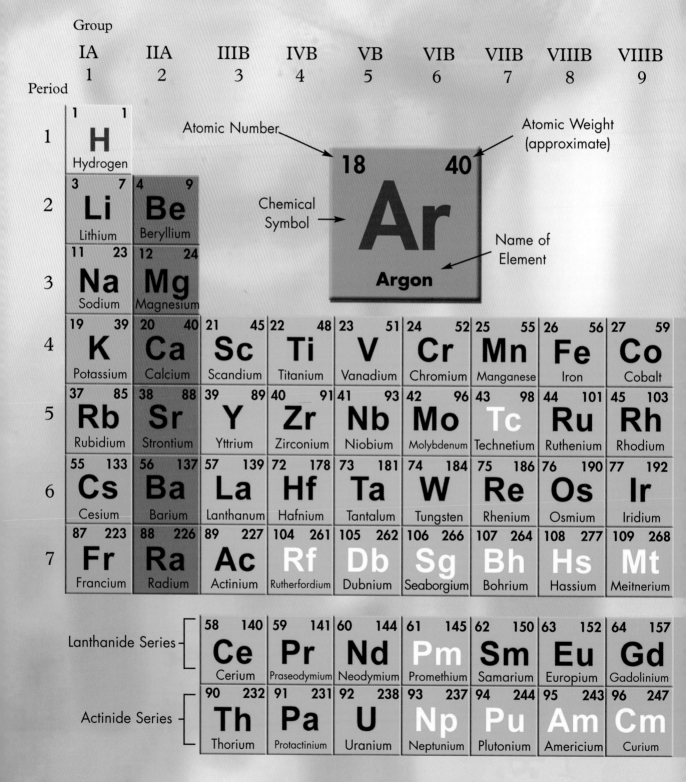

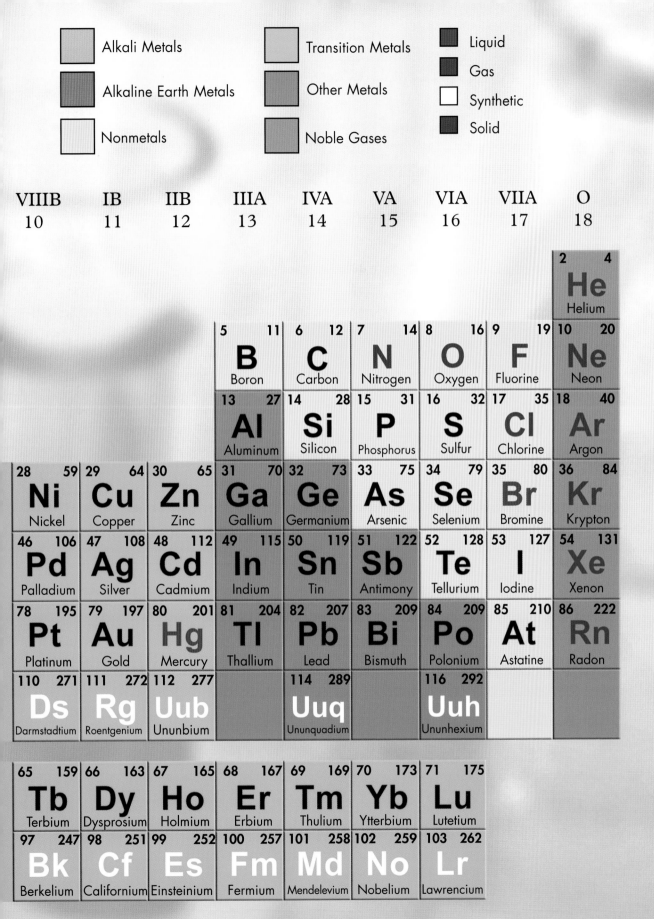

Glossary

alloy A mixture of two or more elements, at least one of which is a metal.

anthropologist A scientist who studies ancient human life and culture.

atomic mass The average mass of all naturally occurring isotopes of an element, taking into account the relative abundance of the isotopes.

atomic number The number of protons in the nucleus of an element.

density A measure of mass (g) per unit volume (cm^3). To find the density of an object, divide the mass of the object by its volume.

electron configuration The arrangement of electrons in atomic orbitals.

emission spectroscopy A technique that analyzes lines of light emitted from excited atoms; it can be used to identify elements, including new ones.

genus A division of organism classification below family and above species, as in *Homo* (the genus that includes *Homo sapiens*, modern humans).

geologist A scientist who studies the origin, history, and structure of the planet, which includes studying rocks.

half-life The amount of time required for half of the radioactive atoms in a sample to decay to another form.

inert Chemically inactive or nonreactive.

ion A positively or negatively charged atom or group of atoms.

ionic bond A chemical bond formed by the attraction between ions with opposite charges.

isotope One of several forms of an element that contains the same number of protons as all other atoms of that element, but a different number of neutrons, resulting in a different mass number.

mass number The total number of protons and neutrons in a nucleus.

octet rule A chemical rule of thumb that says certain atoms will gain, lose, or share electrons in order to fill their outermost energy level with eight electrons.

paleontologist A scientist who studies the fossils of organisms, including plants, animals, and dinosaurs.

radioactive A property of certain isotopes to undergo nuclear reactions (decay), which may cause the change of one element into another element and/or the emission of high energy electromagnetic radiation and/or high-energy particles from their nuclei.

radioactive decay A nuclear reaction of a radioactive atom to produce another element or isotope.

valence electrons Electrons in the highest energy level of an atom.

American Chemical Society

1155 16th Street NW

Washington, DC 20036

(800) 227-5558 (U.S. only)

(202) 872-4600 (outside the U.S.)

Web site: http://www.chemistry.org/kids

The American Chemical Society Web site contains a collection of hands-on activities, games, interactive activities, and articles written for children in the fourth through sixth grades. Print materials can also be obtained from the online store.

Chemical Institute of Canada

130 Slater Street, Suite 550

Ottawa, ON K1P 6E2

Canada

(888) 542-2242

Web site: http://www.cheminst.ca/index.cfm/ci_id/1452/la_id/1.htm

Find out about science fairs, science Olympiad competitions, National Chemistry Week, and other events happening around Canada at CIC's Web site. There are also pages for the junior chemist and engineer that feature articles, experiments, and chemistry-related trivia.

International Union of Pure and Applied Chemistry (IUPAC)

IUPAC Secretariat

104 T. W. Alexander Drive, Building 19

Research Triangle Park, NC 27709

(919) 485-8700

Web site: http://www.iupac.org/dhtml_home.html

IUPAC is an international body designed to advance the chemical sciences.
Science Across the World (http://www.scienceacross.org) and Young
Ambassadors for Chemistry (YAC) are just a few of the educational
programs you will discover on its Web site.

Jefferson Lab

Office of Science Education

628 Hofstadter Road, Suite 6

Newport News, VA 23606

(757) 269-7567

Web site: http://education.jlab.org

Thomas Jefferson National Accelerator Facility (Jefferson Lab) is a
research laboratory that conducts basic research of the atom's
nucleus at the quark level. Its Web site features homework help and
online games and puzzles for students and resources for teachers.

Web Sites

Due to the changing nature of Internet links, Rosen Publishing has
developed an online list of Web sites related to the subject of this book.
This site is updated regularly. Please use this link to access the list:

http://www.rosenlinks.com/uept/argo

Goldenberg, Linda. *Little People and a Lost World: An Anthropological Mystery.* Minneapolis, MN: Twenty-First Century Books, 2007.

Jackson, Tom. *Radioactive Elements.* New York, NY: Marshall Cavendish Benchmark, 2006.

Matthews, John R. *The Light Bulb.* New York, NY: Franklin Watts, 2005.

Panchyk, Richard. *Archaeology for Kids: Uncovering the Mysteries of Our Past.* Chicago, IL: Chicago Review Press, 2001.

Sherman, Josepha. *Henry Cavendish & the Discovery of Hydrogen.* Newark, DE: Mitchell Lane Publishers, 2005.

Thomas, Jens. *Noble Gases.* New York, NY: Benchmark Books, 2003.

Tocci, Salvatore. *Hydrogen and the Noble Gases.* New York, NY: Children's Press, 2004.

Tocci, Salvatore. *The Periodic Table.* New York, NY: Children's Press, 2004.

Trueit, Trudi Strain. *Fossils.* New York, NY: Franklin Watts, 2003.

Willett, Edward. *Neon.* New York, NY: The Rosen Publishing Group, 2006.

Bibliography

Albin, Edward. *Earth Science: Made Simple.* New York, NY: Broadway Books, 2004.

Ball, Philip. *The Ingredients: A Guided Tour of the Elements.* Oxford, England: Oxford University Press, 2002.

Chemical Heritage Foundation. "Gilbert Newton Lewis and Irving Langmuir." September 2006. Retrieved May 16, 2007 (http://www.chemheritage.org/classroom/chemach/chemsynthesis/lewis-langmuir.html).

Dutta, S. K., and A. B. Lele. "Stainless Steel Production by IF-AOD / MRK - LRF Route." December 2004. Retrieved May 16, 2007 (http://jpcindiansteel.org/ssteel.asp).

Emsley, John. *Nature's Building Blocks: An A–Z Guide to the Elements.* New York, NY: Oxford University Press, 2001.

Facts On File News Services. "Six Sure-Fire Methods to Get You a Date." August 2001. Retrieved May 16, 2007 (http://www.2facts.com).

Giunta, Carmen. "The Discovery of Argon: A Case Study in Scientific Method." Le Moyne College, Syracuse, NY. June 2004. Retrieved May 14, 2007 (http://web.lemoyne.edu/~giunta/acspaper.html).

Harris, Tom. "How Light Bulbs Work." HowStuffWorks.com. February 2002. Retrieved May 16, 2007 (http://home.howstuffworks.com/light-bulb.htm).

Holden, Norman. "History of the Origin of the Chemical Elements and Their Discoverers." Brookhaven National Laboratory. March 2004. Retrieved May 14, 2007 (http://www.nndc.bnl.gov/content/elements.html).

International Stainless Steel. "Introduction to Stainless Steel." December 2006. Retrieved May 16, 2007 (http://www.worldstainless.org/About+stainless/What+is/Intro).

Medical Management Services Group. "Laser Eye Surgery & Diseases." Retrieved May 16, 2007 (http://www.seewithlasik.com/docs/laser-eye-surgery-disease.shtml).

Michaels, George H., and Brian M. Fagan. "Potassium-Argon Dating." The University of California at Santa Barbara. December 2005. Retrieved May 13, 2007 (http://id-archserve.ucsb.edu/ANth3/Courseware/Chronology/09_Potassium_Argon_Dating.html).

Nobel Foundation. "Chemistry 1904." Retrieved May 14, 2007 (http://nobelprize.org/nobel_prizes/chemistry/laureates/1904).

PBS. "Saving the National Treasures." NOVA. February 2005. Retrieved May 16, 2007 (http://www.pbs.org/wgbh/nova/charters).

Radiological Society of North America, Inc. "Cryotherapy." Retrieved May 16, 2007 (http://www.radiologyinfo.org/en/info.cfm?pg=cryo&bhcp=1).

Siegel, Lee. "The Oldest Homo Sapiens." University of Utah. February 2005. Retrieved May 13, 2007 (http://www.eurekalert.org/pub_releases/2005-02/uou-toh021105.php).

Stwertka, Albert. *A Guide to the Elements, 2nd ed.* New York, NY: Oxford University Press, 2002.

Thomas Jefferson National Accelerator Facility—Office of Science Education. "Electron Capture." Retrieved May 16, 2007 (http://education.jlab.org/glossary/electroncapture.html).

Thomas Jefferson National Accelerator Facility—Office of Science Education. "The Element Argon." Retrieved May 16, 2007 (http://education.jlab.org/itselemental/ele018.html).

United States Environmental Protection Agency. "Understanding Radiation." November 2006. Retrieved May 16, 2007 (http://www.epa.gov/radiation/understand/index.html).

University of Colorado at Boulder. "Spectral Lines." Retrieved May 14, 2007 (http://www.colorado.edu/physics/2000/quantumzone/index.html).

Index

About the Author

Kristi Lew is a professional K–12 educational writer with degrees in biochemistry and genetics. A former high school science teacher, she specializes in writing textbooks, magazine articles, and nonfiction books about science, health, and the environment for students and teachers.

Photo Credits

Cover, pp. 11, 15, 17, 20, 38–39 Tahara Anderson; p. 5 © John Fleagle, Stony Brook University; pp. 8, 9, 34 © SSPL/The Image Works; p. 10 © ARPLI/Topham/The Image Works; p. 22 © Jon Reader/Science Photo Library/Photo Researchers, Inc.; p. 25 © Phil Mislinski/Getty Images; p. 27 © Fritz Hoffman/Image Works/Time Life Pictures/ Getty Images; p. 30 © Hank Morgan/Photo Researchers, Inc.; p. 32 © Gregor Schuster/Photographer's Choice/Getty Images; p. 33 © Hans Knopf/Pix Inc./Time Life Pictures/Getty Images; p. 35 © Antonia Reeve/Photo Researchers, Inc.; p. 36 © AP Images.

Designer: Tahara Anderson; **Editor:** Kathy Kuhtz Campbell
Photo Researcher: Amy Feinberg